Published by Jastin Enterprises, LLC

Stock Imagery supplied by iStock and Adobe

Any website addresses used in this book are fictitious, however, due to the changing nature of the Internet the web address characteristics may change since publication.



ISBN-13: 978-0692880524

ISBN-10: 0692880526

# Dedication

**BUILDING CYBER-READY WORKERS FROM A YOUNG AGE TO MEET NATIONAL WORKFORCE DEMANDS OF THE FUTURE**

**This book is dedicated to supporting the workforce needs for the 21st century in the areas of cybersecurity. Some surveys estimate that there were over 200,000 cybersecurity jobs left unfilled in 2015 and the demand will grow exponentially over the next 20 years. This book, and subsequent episodes, will educate and inspire a new generation of potential cyber technologists, workers and managers who will have had the opportunity to experience the cybersecurity territory from early childhood thus making "cyber speak" and careers in this area much less foreign.**

**The book targets children between the ages of 8 and 12, as well as adults who like to read with them. Everyone can benefit from reading these episodes in order to become safer online.**

**McGarry (2013) reported that General Keith Alexander, former Director, NSA, described cybersecurity work as a "tremendous opportunity" for young people…" He said - "This generation is coming up cyber savvy," after explaining how his almost 2-year-old granddaughter knows how to use an iPad to watch movies on Netflix. "We can train them. We can educate them."**

*Source: McGarry, B. (Oct. 14, 2013). NSA Chief: What Cyberwarrior Shortage?*

# ACKNOWLEDGMENTS

To all of my friends, family, colleagues and supporters of this effort, I thank you dearly.

~ and ~

To Roy, who fully supported all of my ideas with kindness, respect and endless love.

Jastin is twelve years old and is a very active child on the Internet. In the last episodes about Phishing, Ransomware, Cyber Bullying, and the Internet of Things (IoT), he was still learning about the great adventures with technology and the way to use the new world of "tech toys" – keeping safety first.

Way back in the phishing and ransomware episode, Jastin happily went along posting beautiful pictures online that he found in his house and neighborhood.

He was just trying to have fun on the new website that he joined where his friends and other people chat and share stuff – not knowing that he was giving up valuable personal information.

Sooooo! – here we are in this episode where Jastin learns about **Internet Privacy** and **Identity Theft** which are two main problems that people report to the authorities when it comes to the Internet.

These cybersecurity lessons started one day when Jastin's favorite Auntie Phyl came to visit. Jastin has a huge family and there is something always going on. They all rely on Jastin and ***Super Cee Gee*** for their "tech support" to fix their computers.

Auntie Phyl's story was that she never even uses a computer and brags about that all the time. But, today

Auntie Phyl was so sad. She told Jastin's parents and ***Super Cee Gee*** that someone had emptied out her bank account taking all of her money.

The bank manager told her that someone had **HACKED** into her account along with the accounts of everyone doing business at the bank.

Auntie Phyl needed ***Super Cee Gee*** to explain how that can happen when she would not know a computer if it hit her in the head and never, ever had an online account!

***Super Cee Gee*** let Auntie Phyl know that even if you don't use a computer, the banks use them all the time to handle everyone's account - no matter who they are and what they do online. While there was not much Auntie Phyl could do, the bank was responsible for protecting her money – and they did; they replaced everything that was taken out of her account.

But ***Super Cee Gee*** wanted the family to know that everyone is vulnerable to **HACKERS** and that all of their private information could be stolen, bought, or exposed anytime and possibly used against them.

Listening to Auntie Phyl's story made Jastin and his family think about what's going on with their accounts. In fact, a few days ago they received a weird piece of mail.

***Hmmmmm!!!*** Jastin received a letter from a bank about a credit card that was issued to him. At the time, the family thought it was just a piece of junk mail. But now, having become smarter about this "hacking" stuff from Auntie Phyl's situation, they now knew that someone might have taken Jastin's private information and did something called …

How could someone know so much about Jastin? Stealing a child’s identity is rare. Well he did post a lot of pictures on the friendly-looking website - Remember when Jastin uploaded pictures, names, events and other things to the **www.neighborhoodanimal.org** website?

First, taking pictures in front of the house, told everyone where Jastin lived. Taking pictures of animals under his Dad's car gave everyone information about his Dad's license plate. Of course giving out names, addresses, parents' names and school information told almost the whole world about Jastin's private information. With the **geocodes** on the pictures that Jastin took, a **hacker** could even tell when and where he was when he took them.

Jastin never even thought about looking at the "**Internet Privacy Policy**" on the website to see what they do with his information. And he never thought of asking permission from Mom and Dad because the people on the website seemed so cute and harmless. Awwwwww!

Well, people looking to do bad things on the Internet, posing as friends, could have found out a lot of information about Jastin and family.

So, I guess you ask, why would anyone care? Well, unfortunately not everyone on the Internet is there for good reasons. A lot of times bad people use the internet to lure and kidnap unsuspecting children. Most of these crimes usually start with Internet activity. Just think - it's easy to trick Jastin if someone poses as a family friend who knows where he lives and knows his parent's name from information they got on the internet or from the pictures Jastin posted.

To figure out what this credit card letter to Jastin was all about, the family did some research. They tried to find out what can someone do if they steal your identity? It was mind-boggling for Jastin.

**Hackers** can get into your accounts, like your Instagram, Snapchat or Vine account and post bad things about people or inappropriate pictures. They can get into your gaming system, grab your gamer tags and wipe out all of your game progress scores (Oh my - for Jastin - that's a big deal!).

More importantly, they could also try to illegally get medical assistance under Jastin's name or file a Tax Return that is not real to try to steal money from the U.S. government. *(Jastin doesn't even have a job to pay taxes - **YIKES**!)*

But these hackers wanted quick money and merchandise – So, with personal stuff about Jastin, they were able to open a bank credit card account in his name – yes, a **kid's credit card**? It is possible because they had information about him, his parents, where he lived, and they possibly hacked into school records to get his date of birth and social security number.

Fortunately, the family found out about this credit card fraud in time to take the right steps to stay away from a lot of trouble where someone could have had a mighty good time buying very expensive items with Jastin's fraudulent credit card.

The first thing to do when someone takes control of your identity is to get help – ***QUICKLY!*** - If not, someone could start to make major purchases for luxury items on YOUR account!

Jastin and his parents contacted the bank and the **credit reporting agencies** to let them know about the fraud.

The next lesson was to learn how to keep the family's personal information private and secure when they go to a website – their **personal reputation** is at stake.

Also, there are a lot of ***"Did You Know…"***

***"Did You Know…"*** that the US Government have laws that do not allow websites to collect information from children without their parent's consent?

***"Did You Know…"*** that websites must have what is called an **Internet Privacy Policy** that states what they will do with the child's information? – The policies are usually so long with a lot of legal words – so nobody reads those things!!!

***"Did You Know…"*** that there is a law called

**(COPPA)?**

**Child Online Privacy Protection Act**

specifically created to protect children? Thanks to COPPA Jastin's parents were allowed to delete any information that Jastin put out on the internet because he is under the age of 13.

So for the future, Jastin and Auntie Phyl need to do a few things about protecting their identity in cyberspace.

Well, the first thing they needed to do was to find out exactly what happened by contacting the banks and credit agencies involved in the issue.

Then they had to figure out how to get Jastin's name repaired so that this couldn't happen again. They had to learn how to use the privacy setting on anything that they use online and keep the settings up to date. You can't let bad people be able to figure out where you are by giving them information or pictures online.

And for Auntie Phyl, she can stay away from those "**OLE CONFUSING**" computers, but she has to keep looking at and checking those monthly bank statements that she gets by **snail mail**.

Everyone has to report anything that looks suspicious. !!!!! AND always think before giving out private information!!!!

# Glossary

***COPPA* –** This acronym stands for the ***Children's Online Privacy Protection Act of 1998***, Children's Privacy. This is a federal regulation requiring companies that target their websites or online services to children under the age of 13 to adhere to specific rules. For instance, the website has to disclose what information they are collecting from children and they have to provide special services to allow these children to have their information removed from the online site.

***Credit Reporting Agencies*** – There are three companies that track everyone's credit – that is whether or not you pay your bills on time. Depending on how timely you are in paying your bills, along with some other criteria, the agencies assign credit scores. If your identity is stolen, these agencies can help you monitor and stop the bad activity in your name.

***Cyber Bullying*** – This term is often used to describe many undesirable activities and behaviors related to the Internet, social media, e-mail and texting. The term is derived from traditional bullying in schools, where children were threatened and mistreated in various ways. With so much use of the Internet now, the bullying activities take on a different form but the negative influences and actions are similar and more pervasive. As such, it has become very difficult to monitor and curtail. Everyone must be aware of this phenomenon so that cyber bully victims can be protected and cyber bullies can be taught to respect others.

***Geocoding* –** This is a complicated process of using GPS to match a latitude and longitude to a street address. Geocodes are included in the photos you take and can be used to find someone's location or location of where the photos were taken.

***Hackers and being Hacked* –** There are many kinds of hackers, good ones and bad ones. In the past episodes of Super Cee Gee, the focus has been on the hackers who are trying to get into computers, networks, and systems for bad reasons. Hackers can be one person working out of their home or a number of people in other countries working for large governments. Hackers use their technology to break into someone else's computer, mostly to steal their information (like credit card information and passwords) to sell to others who buy the information illegally. We've learned about some of the ways (phishing and ransomware) that hackers use to get information for disreputable reasons.

***Identity Theft*** **–** Identity theft happens when a hacker (or anyone else) has enough information about you like name, address, social security number, date of birth, school, parents' names, etc. in order to open accounts in your name. Usually they create bank accounts, order credit cards and other accounts that will give them the opportunity to get money – money in your name that YOU may have to pay back. Even if you don't have to pay it back, Identity Theft can ruin your personal reputation. 1-800-IDTHEFT is the number to call for help!

***Internet of Things (IoT)*** - In order to understand IoT, you need to know that computers communicate through the use of "addresses." Every internet device has to have this address to be located by another computer. This "Internet Protocol" (IP) addressing scheme is kind of complicated and you can learn more about that later. Today, there are many, many Internet addresses that we can associate with cars, refrigerators, alarm systems, mobile devices, light bulbs, thermostats and other things. Since these "things" can now communicate over the Internet – this term - IoT was coined. IoT is a good thing, and we easily buy and install devices because of the conveniences they provide - but with so many devices talking to each other and possibly sharing personal information, we need to think about some of the cybersecurity dangers.

***Internet Privacy*** – Maintaining your privacy when you are online is a difficult thing to manage. Many people, young and old, freely put personal information in their social media accounts and give it online to companies from which we buy things. Also, as outlined in the Super Cee Gee "IoT" episode, some information is automatically collected by the devices we use. Online users have to be vigilant about protecting their personally identifiable information (PII) all the time.

***Internet Privacy Policy*** – An internet privacy policy is something that every website is responsible for having. It usually tells you exactly what the company will do with the private information they collect about you. It also states the safety precautions the company takes with your information. But if you are not comfortable with how they protect your data, the policy states how you can get in touch with them to complain.

***Personal Reputation*** – Your personal reputation is what you carry with you throughout your life. In the digital world, it includes all of the information that exists on the Internet about you: pictures, friends, clubs, school, etc. It includes things that you posted and things that someone else has posted about you. It is important that you only have good things associated with your name and it starts with protecting your privacy and your online personality. As you get older, schools, colleges, and employers will be able to look at your online personal reputation.

***Phishing*** – As a reminder from Jastin's first episode, Phishing attacks are named that because they are just like real fishing where someone throws out the bait on a fishing pole to catch a poor, unsuspecting fish and the fish bites the bait and is reeled in? Well in the digital world hackers often send a bogus email to someone (the bait) hoping that some poor "schmuck" will think it's real and send their precious personal information to them. **Wrong**!

***Ransomware*** – Bad software or (malware) that can stop people from accessing their pc, laptop, tablet or smart phone in essence putting a lock on files, pictures, data, contacts, and screens until a ransom is paid to the hacker.

***Snail Mail*** – The letters that you and your family receive from the post office delivered by a mail carrier. Called snail mail in this new digital world because it is so much slower than e-mail and text messages.

Made in the USA
Middletown, DE
14 October 2022